What's for

Corn

Library of Congress Cataloging-in-Publication Data
Robson, Pam
  Sweetcorn / written by Pam Robson. -- 1st American ed.
  p.   cm. -- (What's For Lunch?)
  Included index.
  Summary: An introduction to the corn we eat, starting at the farm
with its planting and harvesting, until it is sold in markets. Also
discusses the manufacture of other products made from corn.
  ISBN 0-516-20823-3
  1. Sweet corn--Juvenile literature. 2. Corn products--Juvenile
literature.   [1. Corn.]   I. Title.   II. Series: Robson, Pam.
What's For Lunch?
SB351.C7R63   1998
635'.672--dc21   97-6910
                                              CIP
                                              AC

© 1997 Franklin Watts
96 Leonard Street
London
EC2A 4RH

First American edition 1998 by
Franklin Watts
A Division of Grolier Publishing
Sherman Turnpike
Danbury, CT 06816

ISBN 0-516-20823-3 (lib. bdg.)
ISBN 0-516-26219 X (pbk)

Editor: Samantha Armstrong
Series designer: Kirstie Billingham
Designer: Dalia Hartman
Consultant: Anne Stone, Phd.
Reading Consultant: Prue Goodwin, Reading and Language
Information Centre, Reading

Printed in Hong Kong

What's for lunch?

Corn

Pam Robson

CHILDREN'S PRESS®

A Division of Grolier Publishing

LONDON • NEW YORK • HONG KONG • SYDNEY
DANBURY, CONNECTICUT

Today we are having corn for lunch.

Corn is a vegetable.

Sometimes we eat corn
as corn on the cob.

Corn contains sugar and **starch**.

Eating corn will give you **energy**.

Corn comes from a plant called maize.
Maize can be very colorful.
Some maize is green, brown, red, or purple.
Maize is also used to make popcorn,
cornflakes, and animal food.

Maize grows in many different countries.
Some kinds of maize grow best
in hot places like Brazil, in South America.

Other kinds grow best in cooler places,
like Nebraska, in the United States.
Most of the maize we eat comes from
the United States.
In America, maize is called corn.

Like all plants,
maize must have water
and sunshine to grow.
In hot countries,
maize seeds are planted
when there is plenty of rain.
In cooler countries,
the seeds are planted
in the spring.
A large machine
sows the seeds in the soil.

Farmers put **fertilizer** on the soil to help the maize grow strong. Insects sometimes eat the growing plants. Airplanes spray the **crop** with **insecticide** to kill the insects.

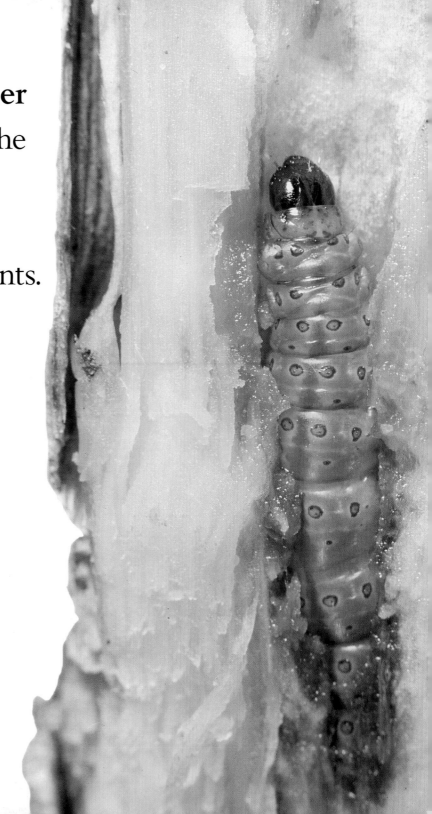

Maize has both male and female flowers.
The wind blows **pollen** from the
male flowers to the female flowers.
The female flowers have threads
called **silks** to collect the pollen.
The pollen makes seeds grow
inside the female flowers.
The female flowers are covered with
special leaves called **husks** to protect them.

The seeds are called **kernels**.
The kernels grow on a **cob**.
Some types of corn have soft,
sweet kernels.
When the kernels are yellow,
a machine snaps the cobs
off the plant.
When we eat corn
at this stage, it is called
corn on the cob.

Sometimes we take the kernels off the cob.
We call the kernels corn.

Corn usually tastes best when it is fresh.
So it is quickly packed into cans or it is frozen.

Sometimes corn is treated
in a different way.
A machine called a
harvester picks the cobs
from the plants.
Then the husks
are removed and
the kernels are taken out.
We call this **threshing**.

The corn goes to mills where machines clean it, dry it, and grind it. The ground corn is made into animal food, glue, or cornstarch.
It can be finely ground into **cornmeal** to use in cooking.
It is also used to make cornflakes.

glue

Corn products often travel
many miles before being eaten.
Cans of corn and boxes of cornstarch
are packed on trains and ships and planes.
When they arrive, they are taken to the shops.
There they are put on shelves for us to buy.

Many different products
come from corn.
Corn oil is used for cooking.
Popcorn comes from kernels
that have been dried
and are heated up until they pop.
Corn chips are crunchy.

Maize is eaten all around the world.
In Mexico, cornmeal is used to make
crispy pancakes called taco shells.
Delicious fillings are put inside them.

In America corn muffins
are often eaten for breakfast.
Maize is a very useful plant.

Glossary

cob   the hard center of corn on which the seeds, or kernels, of corn grow

cornmeal   a yellow flour made from crushed grains of corn

cornstarch   a smooth white flour made from ground corn

crop   what farmers grow in their fields

energy   the strength to work and play

fertilizer   something that helps plants to grow

harvester   a large machine that cuts down the maize plant and removes the cobs from the plant

husk	the outside covering of a fruit or seed
insecticide	something that kills insects
kernel	a seed of corn inside the husk
pollen	powder made by male flowers that fertilizes the female flowers
silk	a tuft growing on the female flower of the maize plant
starch	a white substance found in certain foods such as potatoes, rice, and corn
threshing	removing the husks and separating the grains of corn from the husk
vegetable	a plant grown for the parts that can be eaten

Index

Picture credits: Eye Ubiquitous 12 (Adrian Carroll); By kind permission of Green Giant ® 18, 19;
Holt Studios International 6 (Inga Spence), 10-11 (Willem Harinck), 13 (Len McLeod), 14 (Nigel Cattlin),
15 (Nigel Cattlin), 16-7 (Nigel Cattlin), 20-1 (Willem Harinck), 24 (Inga Spence); Images 7, 8-9;
Nick Bailey Photography cover, 3, 5. All other photographs Tim Ridley, Wells Street Studio, London.
With thanks to Lois Browne and Ushil Patel.